Ocean Babies

Mary Elizabeth Salzmann

Consulting Editor, Diane Craig, M.A./Reading Specialist

Sandcastle

An Imprint of Abdo Publishing
abdobooks.com

abdobooks.com

Printed in the United States of America, North Mankato, Minnesota

052019
092019

Design: Christa Schneider, Mighty Media, Inc.
Production: Mighty Media, Inc.
Cover Photograph: Shutterstock Images
Interior Photographs: Shutterstock Images (all)

Library of Congress Control Number: 2018966953

Publisher's Cataloging-in-Publication Data
Names: Salzmann, Mary Elizabeth, author.
Title: Ocean Babies / by Mary Elizabeth Salzmann
Description: Minneapolis, Minnesota : Abdo Publishing, 2020 | Series: Animal babies
Identifiers: ISBN 9781532119606 (lib. bdg.) | ISBN 9781532174360 (ebook)
Subjects: LCSH: Marine animals--Juvenile literature. | Animal babies--Juvenile literature. | Sea animals--Juvenile literature.
Classification: DDC 591.3909162--dc23

SandCastle™ Level: Emerging

SandCastle™ books are created by a team of professional educators, reading specialists, and content developers around five essential components—phonemic awareness, phonics, vocabulary, text comprehension, and fluency—to assist young readers as they develop reading skills and strategies and increase their general knowledge. All books are written, reviewed, and leveled for guided reading and early reading intervention programs for use in shared, guided, and independent reading and writing activities to support a balanced approach to literacy instruction. The SandCastle™ series has four levels that correspond to early literacy development. The levels are provided to help teachers and parents select appropriate books for young readers.

EMERGING • BEGINNING • TRANSITIONAL • FLUENT

Contents

Ocean Babies

Can you find these baby ocean animals in this book?

baby dolphin

baby eel

baby sea otter

baby
sea turtle

baby
seahorse

baby seal

baby shark

baby whale

The baby dolphin can swim.

The baby sea turtle can swim.

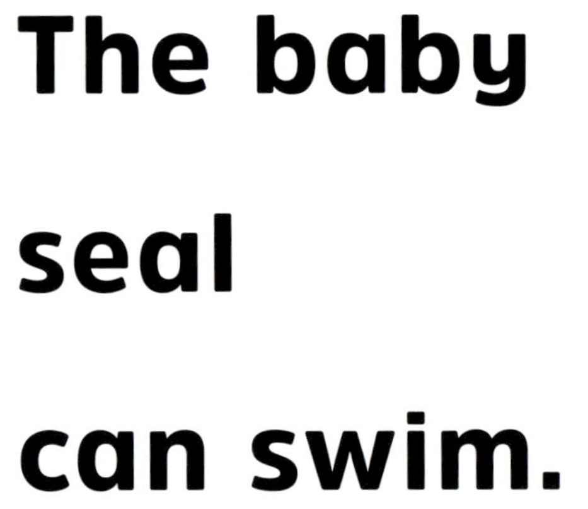

The baby seal can swim.

The baby whale can swim.

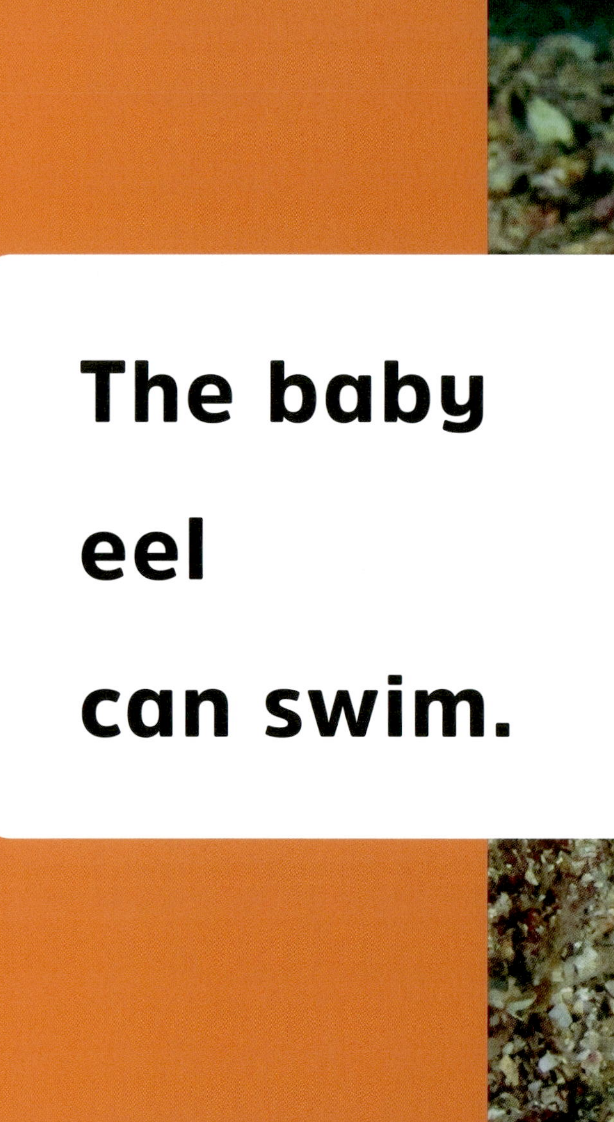

The baby
eel
can swim.

The baby seahorse can swim.

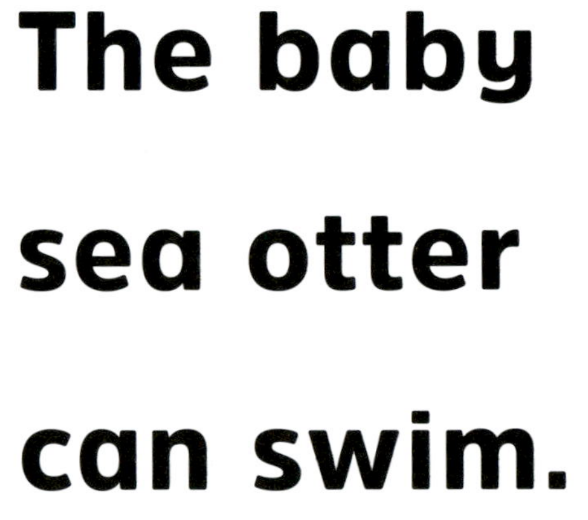

The baby sea otter can swim.

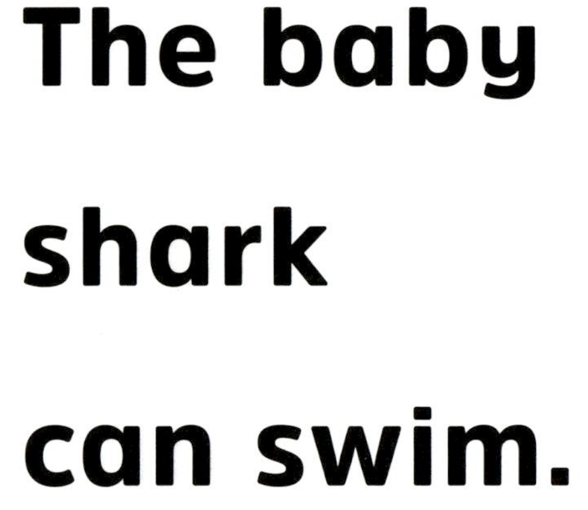

The baby shark can swim.

What Else Did You See?

coral

plants

sand

water

Index

Teacher's Guide

ATOS: 1.0 GRL: A Word Count: 42

High-Frequency Words

can The

Content Words

baby, dolphin, eel, sea otter, sea turtle, seahorse, seal, shark, swim, whale

Before Reading

- Tell students that the title of the book is *Ocean Babies.*
- Summarize the content of the book.
- Have students look through the book. Ask them what they see in the pictures.
- Choose a few new vocabulary words. Have students predict what letter each word starts with. Then have them find the words in the book.

After Reading

Ask students questions about the book's content, such as:

- What animals did you see in the book?
- How could you tell that the animals were in the ocean?
- Have you ever been to the ocean? What was it like?
- What other animals would you like to read about?